# A Monarch Adventure

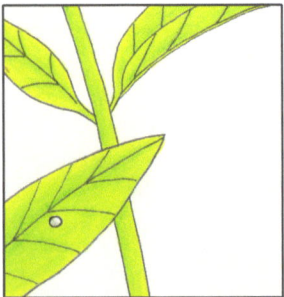

## A Word About New Words

At the end of this book is a list of some nature vocabulary used in the story that might be new to you. While you're reading, if you come across a word you don't recognize, look for it in the glossary to learn something new!

In loving memory of Chris, gone from sight but forever in our hearts.  - BL

To my parents (Beth's grandparents), Nancy and Bill Fulton, who encouraged us all by their example, to read every book we can, to explore our creativity, and to take a walk in the woods every chance we get.  - MK

A Monarch Adventure

Text and photographs © 2021 by Beth Lawnicki
Flip book illustrations © 2021 by Mary Katanik
Nature journal illustrations (pp. 24-25) © 2021 by Mia, Lily, and Nico Lawnicki
Photo of butterfly milkweed (page 9) by Chris Lawnicki in 2016

ISBN 978-1-7367055-0-6

The sunlight caught a flash of orange.

Lia was walking home from the park when she saw a butterfly floating by on the way to a flower.

"What kind of butterfly is that?"

When she got home, seven-year-old Lia researched butterflies and found the one she had seen - a monarch (say it: mon-arc).

She discovered monarchs start life as an egg, and then hatch into a tiny caterpillar, striped in yellow, white, and black.

Then they
eat,
eat,
eat, to
grow,
grow,
GROW

By the time it's done, that tiny caterpillar will have grown to 3,000 times its original size! Once a monarch caterpillar is big enough, it changes into a butterfly. Lia was amazed by the transformation she read about. She wanted to learn more.

Lia learned that monarch butterflies go on an amazing journey. They migrate south each year during autumn in North America, to spend winter in warmer areas.

In the spring, monarchs return north.

The number of monarch butterflies in the world is shrinking because their habitat and food sources are disappearing.

Lia read about milkweed, the "host plant" for monarchs.  As people landscape, farm, and use chemicals to spray for weeds, there is less milkweed for the monarchs.

But milkweed is the ONLY plant monarchs lay eggs on, and the only plant monarch caterpillars eat!

Lia went to the plant nursery and chose some milkweed plants with beautiful flowers.

She planted most in the garden, but saved one milkweed to place inside her screened-in patio.

Then Lia waited . . .

Every day Lia went to the garden and checked each milkweed leaf to see if she could find a tiny egg. And one day she did! Soon, the egg hatched and Lia found a teeny, tiny, itty-bitty caterpillar, so small that she almost couldn't see it. She named it "Bub."

She carefully brought Bub and his leaf to the milkweed plant on her patio.

Each day she checked on him, as soon as she woke in the morning and again when she walked in the door after school.

That tiny caterpillar munched and munched, and he grew and grew. Some days she felt like she could see how much bigger he was just since the last time she'd checked.

As he grew, Bub became too big for his hard outer layer. He shed this exoskeleton (say it: ek-soh-skel-uh-tin) five times to make room to keep growing.

After a week and a half of munching and growing, Bub was the size of Lia's index finger.

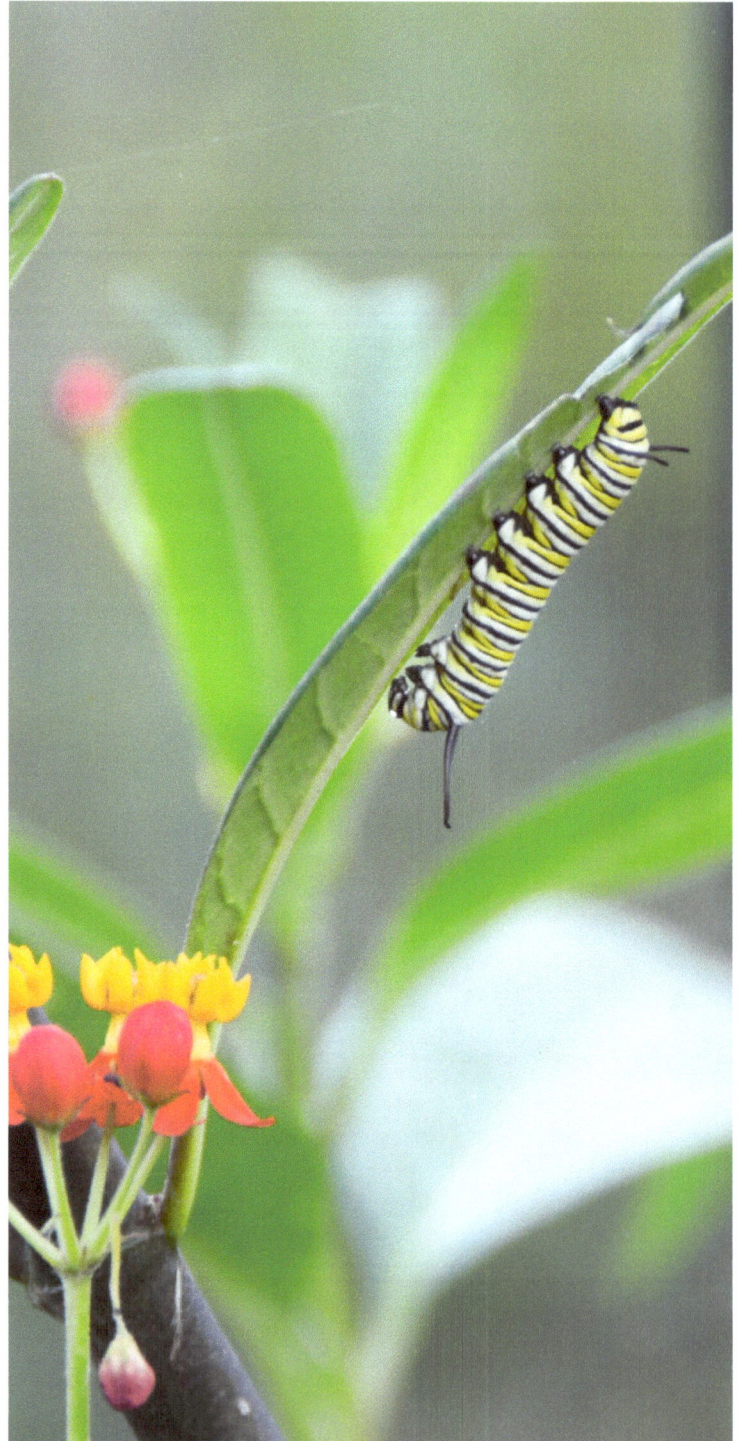

One day, Bub crawled down from the plant and up the inside of the patio screen.  He found a sheltered area.

That night, Lia found him hanging upside down in a "J" shape. And, the next morning, Bub did something amazing.

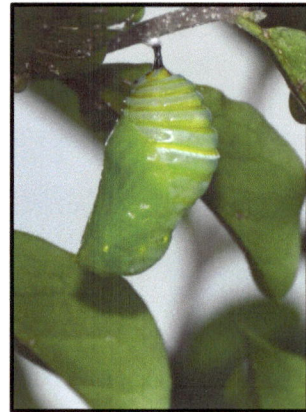

Lia watched Bub form a chrysalis (say it: cri-suh-luss), starting with the bottom of the "J" and moving to the top. He wiggled and wriggled for a minute or two until he was completely hidden inside a mint green chrysalis.

Nine days later, Lia noticed that the green chrysalis was becoming darker. By the next morning, it had turned black and she could see some orange coloring right through the outer layer. Can you guess what happened next?

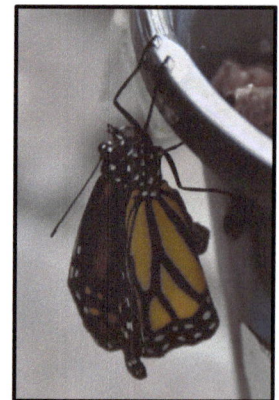

As Lia watched, the chrysalis split open and a creature slowly emerged.  It had a small head, a large abdomen and wet, crumpled wings.  It didn't look anything like a caterpillar.

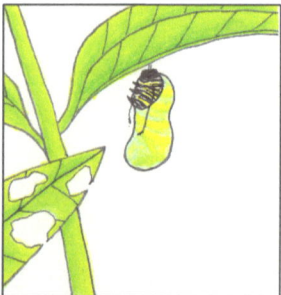

18

Slowly, Bub unfurled his wings, and he was a beautiful monarch butterfly - just like the one Lia had seen near the park!

Bub the butterfly started taking his first steps. He flapped his wings to expand and dry them. Soon he began flying.

Lia released Bub from the patio, and he flew a short way to a tree in her yard.

Then he flew up and away until she couldn't see him anymore. Happy flying, Bub!

Watching Bub grow taught Lia so much about the life cycle of monarchs. But she had many more questions.

Lia found more monarch eggs on her milkweed. She watched their growth and transformations with curiosity.

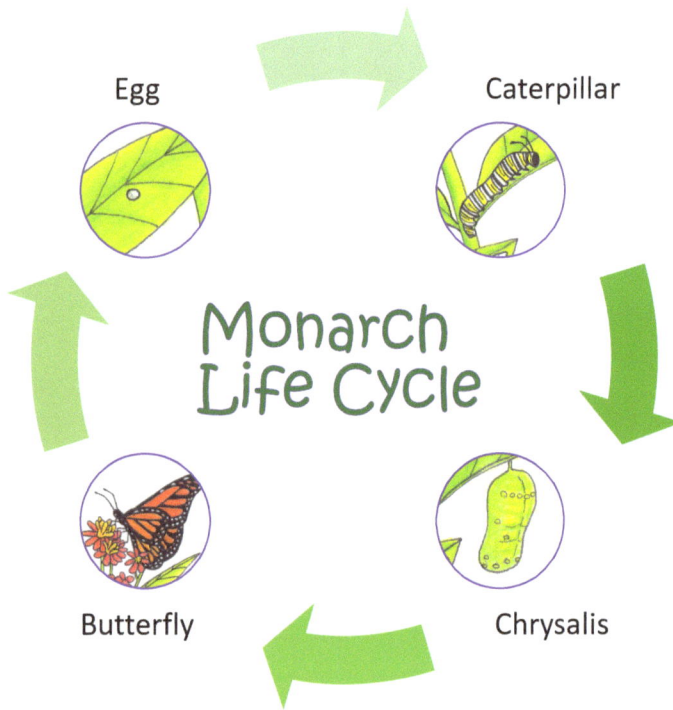

Egg

Caterpillar

Monarch
Life Cycle

Butterfly

Chrysalis

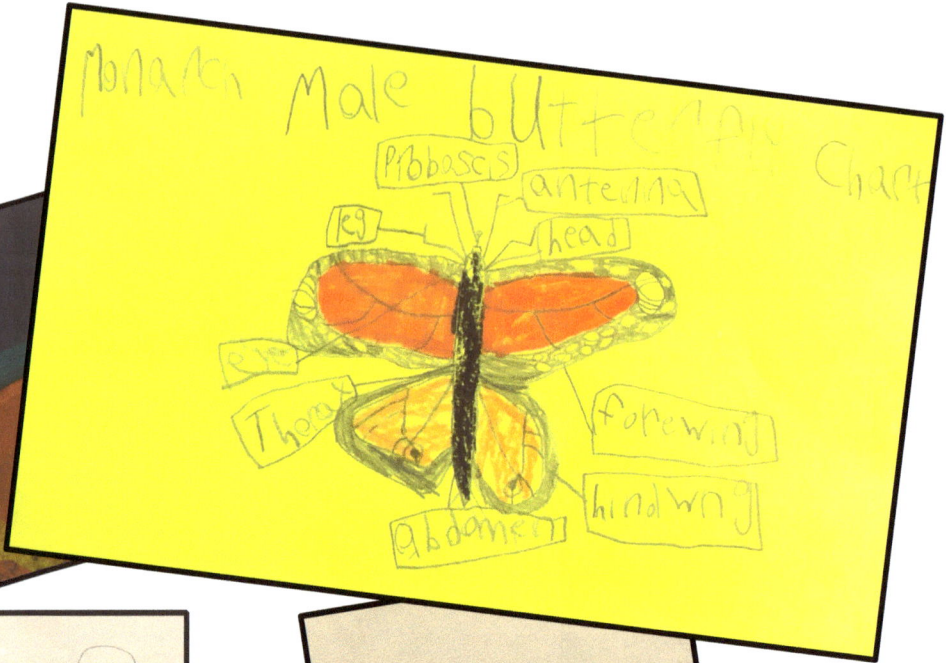

Monarch Male Butterfly Chart

Proboscis
antenna
leg
head
eye
Thorax
forewing
hindwing
Abdomen

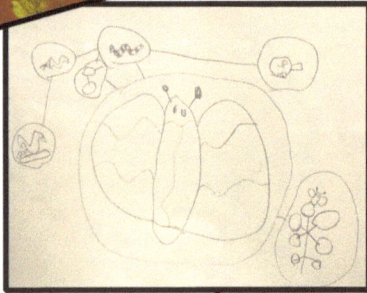

Monarch

Lia kept a nature journal of her observations, including notes, photographs, names and drawings.  She listed the questions she had and wrote answers she found.

She shared her journal with her family and friends, and showed a video of the monarch transformation to her class at school.

Lia hoped that sharing her experience with others . . .

Butterfly

Egg

Caron

exoskelton shedn

Baby caterpilan

How to draw a Butterfly!

. . . would raise
awareness and help
save the monarchs.

# How can YOU help the monarchs?

• Be curious and learn about monarchs, their habitat, food sources, predators, life cycle and migration. Learn how they're unique and why they're important. Online resources and books have so much information about monarchs.

• Build a butterfly garden by planting native nectar flowers and milkweed in your yard or community (where you're allowed!).

• Visit a local botanic garden or arboretum.

• Ask your community garden, school, or class to work on a monarch project so others can learn about monarchs, too. Spread the word in your community and school about the monarchs and the help they need for their species to survive.

• Use your imagination to think of ideas or projects to help. Kids have a unique perspective and can think of so many creative ways to help protect nature!

# Fun Facts About Monarchs

1. Monarch butterflies taste with their feet and drink nectar from flowers using their proboscis.

2. Monarchs have four wings.

3. Monarchs are insects and have four stages of growth: egg, larva (caterpillar), pupa (chrysalis), and adult (butterfly).

4. Monarch caterpillars shed their exoskeleton (called molting) five times, and the stages between molts are called instars.

5. Monarch caterpillars make a very strong silk pad to attach to when forming a chrysalis. Look at the photos on page 18. The tiny white area at the top of the J is the silk pad.

6. Monarch butterflies usually live from 2 to 5 weeks, but the generation of monarchs that migrate south can live as long as 9 months!

7. A female butterfly can lay over 1,000 eggs during her life. In the wild, usually less than 10% (under 100) of the eggs survive to become butterflies.

8. Monarch butterflies need warm temperatures to fly, at least 55 degrees Fahrenheit (13 degrees Celsius).

9. Because milkweed sap (the juice inside the plant) is poisonous, and monarchs eat it, monarch caterpillars and butterflies are poisonous if eaten. This is a defense mechanism to protect them from predators.

# Author's Note

The story of the monarch is a story of transformation, from a land-bound, slow, crawly caterpillar to a graceful, gliding creature. In the summer of 2020, in the midst of months of stay-at-home orders and social distancing due to COVID-19, my three children (Lily, Mia and Nico) and I planted milkweed in our backyard "for the butterflies." We were ready for something new to do, to grow, to learn. After observing several monarchs from egg through butterfly, we were captivated. We planted more milkweed, and more caterpillars arrived. There were several emergency trips to the garden center to get more milkweed when the caterpillars had decimated our current crop. We loved to watch butterflies in our yard and each sighting was as exciting as the first one.

Earlier in the year we had lost someone very dear to our hearts - my husband and the kids' dad - to cancer. Many cultures associate butterflies with the transformation from life (caterpillar) through death (chrysalis) to an afterlife free from the constraints of the earth-bound form (butterfly). I think without realizing it, we were drawn to butterflies that summer because we felt a connection through them to the kids' dad. Seeing that transformation over and over helped to ease some of our grief, and was a reminder that now he was free like the butterflies.

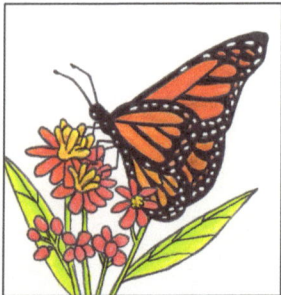

*Flip it!* Hold all the pages in one hand and then let them go one by one, starting with page 2, while looking at the images in the boxes. If you do it fast enough, it will look like the monarch is growing!

# Glossary

**Chrysalis (say it: cri-suh-luss):** the hard outer shell protecting the insect as it transforms from a caterpillar into a butterfly.

**Exoskeleton (say it: ek-soh-skel-uh-tin):** the hard outside skeleton of the caterpillar that protects and supports it. Like the internal skeleton that people have but on the outside!

**Habitat:** the place that an animal lives, where it finds shelter, food and water.

**Host plant:** the plant that butterflies lay their eggs on and need for their caterpillar form to eat and grow. For monarchs, the host plant is milkweed.

**Migrate:** to move from one region to another. Monarchs migrate south to Southern California or Mexico for the winter.

**Milkweed:** the host plant for monarchs. Monarch caterpillars only eat milkweed plants.

**Native plants:** plants that grow naturally in a particular region without being introduced by humans. Over time insects have evolved and adapted to native plants.

**Nature journal:** a place that you can record what you see and learn about the natural world. A nature journal can include things like drawings, diagrams, poems, photographs, stories and observations. Whatever helps you to understand the world around you can go into a nature journal.

**Nectar:** a sweet liquid in some flowers that attracts insects. Butterflies love to drink nectar.

**Predator:** an animal that hunts and eats other animals. Monarch predators include wasps, ants and birds.

**Proboscis (say it: proh-boss-kiss):** a long, tube-like and flexible mouth that a butterfly uses to sip nectar.

www.ingramcontent.com/pod-product-compliance
Lightning Source LLC
Chambersburg PA
CBHW061153030426
42336CB00002B/30